AF601289

Chasing the CHUPACABRA

Anna Anderhagen

Big Buddy Books
An Imprint of Abdo Publishing
abdobooks.com

abdobooks.com

Published by Abdo Publishing, a division of ABDO, PO Box 398166, Minneapolis, Minnesota 55439.

Printed in the United States of America, North Mankato, Minnesota
102023
012024

Design: Denise Hamernik, Mighty Media, Inc.
Production: Mighty Media, Inc.
Editor: Liz Salzmann
Cover Photograph: Karrrtinki/Adobe Stock
Interior Photographs: Alexlky/Shutterstock Images, p. 5; Casey Bisson/Flickr, p. 27; Daniel Eskridge/Adobe Stock, p. 7; Eric Gay/AP2007, p. 19; FlixPix/Alamy Photo, p. 29; Franco Folini/Wikimedia Commons, p. 25; Kit Leong/Shutterstock Images, p. 17; Michael Snipes/Flickr, p. 13; Sgerbic/Wikimedia Commons, p. 21; Shutterstock Images, p. 9; United Archives GmbH/Alamy Photo, p. 11; Worldillustrator/Adobe Stock, p. 15
Design Elements: Joko/Adobe Stock (tape and paper); matiasdelcarmine/Adobe Stock (silhouette); Melica/Adobe Stock (polaroid frame); Net Vector/Shutterstock Images (map); Oleg Iatsun/Shutterstock Images (compass); Piman Khrutmuang/Adobe Stock (paper); STILLFX/Shutterstock Images (background texture); Wandeaw/Shutterstock Images (background texture); Yevhenii/Adobe Stock (tape); YuliaZelinskaya/Shutterstock Images (footprints)

Library of Congress Control Number: 2023939277

Publisher's Cataloging-in-Publication Data
Names: Anderhagen, Anna, author.
Title: Chasing the chupacabra / by Anna Anderhagen
Description: Minneapolis, Minnesota : Abdo Publishing, 2024 | Series: Chasing cryptids | Includes online resources and index.
Identifiers: ISBN 9781098291891 (lib. bdg.) | ISBN 9781098278793 (ebook)
Subjects: LCSH: Chupacabras--Juvenile literature. | Monsters--Juvenile literature. | Animals, Mythical--Juvenile literature. | Folklore--Juvenile literature. | Cryptozoology--Juvenile literature.
Classification: DDC 001.944--dc23

CONTENTS

CHAPTER 1

CHASING CRYPTIDS

Did you hear that **slurping** sound? You and your friend stop on the dirt road. You see a creature **hunched** over a dead animal. It looks at you with red eyes. You see its **fangs**. Suddenly, it hops away like a kangaroo! Could this be the Chupacabra cryptid that you have been looking for?

Chupacabra means "goat sucker" in Spanish. People who believe in the Chupacabra say it is like a vampire. It drinks the blood of its victims.

CHAPTER 2

WHAT IS A CRYPTID?

A cryptid is an animal not proven to exist. There are stories about many different cryptids. A well-known cryptid is the Chupacabra. Many people think they have seen a Chupacabra, but no one has been able to prove it. People **describe** the Chupacabra in many ways, depending on where they live. And the description of the Chupacabra has changed over time.

Some people say the Chupacabra has sharp spikes on its back.

CHAPTER 3

DESCRIPTIONS OF THE CHUPACABRA

The first Chupacabras were said to have long claws, **fangs**, and huge red eyes. They were 4 to 5 feet (1.2 to 1.5 m) tall, walked on two feet, and could jump like a kangaroo.

Later, people said Chupacabras were much smaller. They walked on four legs and had blue skin. They looked like hairless wolves or dogs.

CRYPTID PROFILE

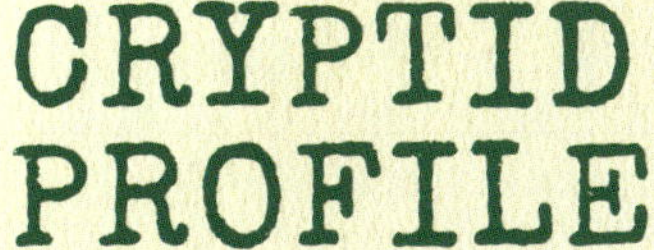

NAME: Chupacabra

CLASSIFICATION: mammal

LOCATIONS: Puerto Rico, United States, Mexico, Chile, Brazil

HABITATS: grasslands, forests

DESCRIPTION:

— long claws
— fangs
— red eyes
— blue skin
— spikes

PROVEN TO EXIST: not yet

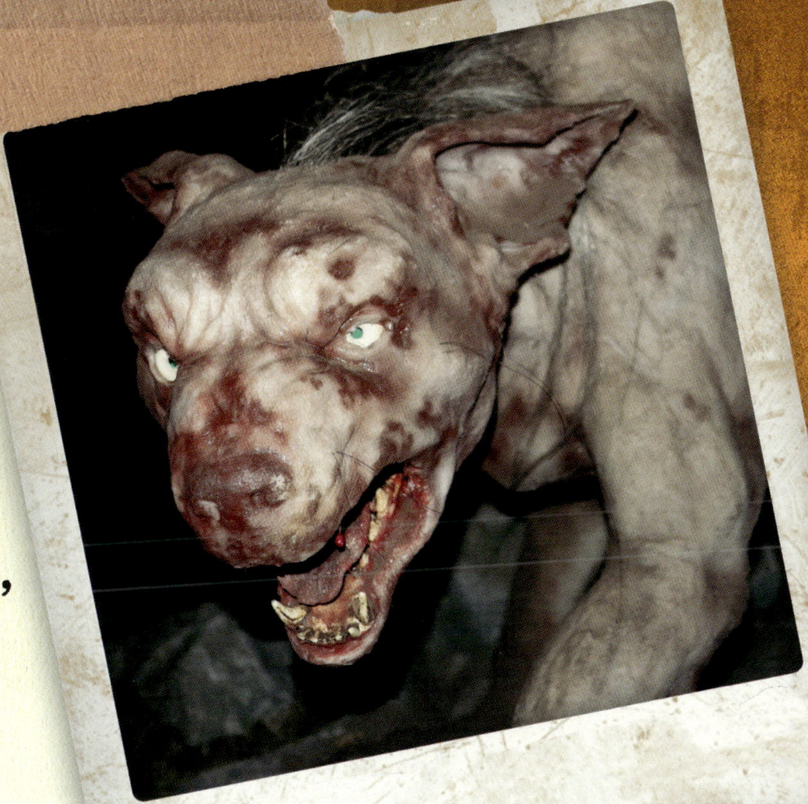

CHAPTER 4

EARLY SIGHTINGS

The Chupacabra was first seen in Canovanas, Puerto Rico, in 1995. The first person to report seeing it was Madelyne Tolentino. She saw it through her window.

The creature had huge **protruding** eyes, long arms, and spikes down its back. It had purple skin and long legs. Tolentino screamed and the creature hopped away like a kangaroo.

Some researchers think Tolentino's sighting was influenced by a movie she saw. Her description of the Chupacabra was like the alien in the movie *Species*.

That same year, farmers in Canovanas blamed the Chupacabra for killing their livestock. They said it was a creature with **fangs** and huge eyes. The dead animals had teeth marks on their throats.

In the town of Canovanas, 150 animals were reportedly killed by the Chupacabra.

Misael Negron was a college student living in Canovanas. He said he saw the Chupacabra suck blood from a goat's neck. Negron **described** the Chupacabra as having skin like a dinosaur and multicolored spikes down its head and back.

Negron said the Chupacabra had red eyes the size of chicken eggs.

CHAPTER 5

THE ELMENDORF BEAST

Devin McAnally was a rancher in Elmendorf, Texas. In 2004, something started killing his chickens. Then, he saw a creature with **fangs** and blue skin eating berries under a tree. He shot and killed the beast. Some people thought it was a hairless dog or a sick deer. Others thought it was a Chupacabra.

Elmendorf is near San Antonio, Texas. McAnally showed his beast to experts at the San Antonio Zoo. They were not able to identify it.

CHAPTER 6

HAIRLESS DOG?

In 2007, Phylis Canion saw a strange creature on her ranch in Cuero, Texas. She **described** it as a hairless dog with blue skin. Over the next few days, she found many of her chickens dead. The chickens had teeth marks on their necks. Later, she found the creature dead by the side of the road. She thinks it's a Chupacabra.

Canion with the head of the creature she found

CHAPTER 7

CHUPACABRA INVESTIGATION

Benjamin Radford spent five years in the early 2000s **investigating** the Chupacabra. He listened to people's Chupacabra stories. He even **examined** Canion's creature. Radford thinks that what people believe are Chupacabras are actually sick coyotes or dogs.

Radford wrote a book about his investigation called *Tracking the Chupacabra*. It was published in 2011.

CHUPACABRA SIGHTINGS
NORTH AMERICA
SOUTH AMERICA
OREGON
MICHIGAN
ARIZONA
NEW MEXICO
ILLINOIS
NEW JERSEY
TEXAS
FLORIDA
MEXICO
PUERTO RICO
BRAZIL
CHILE
N
W
E
S

The Chupacabra sightings in the 1990s were in Puerto Rico. Most of the sightings since 2000 have been in Texas and several other US states. There have also been sightings in Chile, Brazil, and Mexico.

CHAPTER 8

STUDYING THE CHUPACABRA

Dr. Barry O'Connor is a biologist. He thinks Chupacabras are sick coyotes. He said coyotes can get **infected** with a **parasite**. This parasite makes them lose their fur. O'Connor noted that coyotes infected with the parasite also grow thicker skin and smell terrible.

It's possible that sick coyotes attack livestock because they aren't strong enough to hunt wild prey.

CHAPTER 9

CRYPTID KEEPER: LOREN COLEMAN

Loren Coleman also believes that recent Chupacabras are sick coyotes. But he's not sure the ones in Puerto Rico in 1995 were. The **descriptions** are too different. He thinks they could have been monkeys.

At that time, a Puerto Rican lab was experimenting on monkeys. Coleman thinks that some monkeys may have escaped and were living in the wild.

In 2003, Coleman opened the International Cryptozoology Museum in Portland, Maine.

CHAPTER 10

DO CHUPACABRAS EXIST?

Unlike most cryptids, there are bodies of dead Chupacabras that scientists can study. Almost all the results conclude that the animals are sick coyotes or dogs. But this **evidence** does not explain the larger Chupacabras seen in Puerto Rico in the 1990s. Still, most scientists do not think Chupacabras exist. What do you think?

The 2023 Netflix movie *Chupa* features a baby Chupacabra.

describe—to tell about something with words. Such a telling is a description.

evidence—facts that prove something is true.

examine—to look at closely.

fangs—long, sharp teeth.

hunch—to bend one's body into an arch or hump.

infected—made sick by something harmful entering the body.

investigate—to gather information about or study something. An act of investigating is an investigation.

parasite (PEHR-uh-site)—a living thing that lives in or on another living thing. It gains from its host, which it usually hurts.

protrude—to stick out.

slurp—to make a sucking noise while eating or drinking.

ONLINE RESOURCES

To learn more about the Chupacabra, please visit **abdobooklinks.com** or scan this QR code. These links are routinely monitored and updated to provide the most current information available.

INDEX